5th International Conference on Epigenetics and Bioengineering (EpiBio 2021)

Online
4-6 November 2021

ISBN: 978-1-7138-4438-9

Printed from e-media with permission by:

Curran Associates, Inc.
57 Morehouse Lane
Red Hook, NY 12571

AIChE
120 Wall Street, FL 23
New York, NY 10005-4020

Phone: (800) 242-4363
Fax: (203) 775-5177

www.aiche.org

Additional copies of this publication are available from:

Curran Associates, Inc.
57 Morehouse Lane
Red Hook, NY 12571 USA
Phone: 845-758-0400
Fax: 845-758-2633
Email: curran@proceedings.com
Web: www.proceedings.com

TABLE OF CONTENTS

BIOPHYSICS OF CHROMATIN: MEASUREMENTS AND MODELS

EMERGING TECHNOLOGIES

EPIGENETICS IN HEALTH AND DEVELOPMENT

NUCLEAR ORGANIZATION

ORAL SESSION

PERTURBING CHROMATIN AND EPIGENETIC EDITING

SENSING EPIGENETIC MODIFICATIONS

POSTER SESSION

3D Dynamics of Heterochromatin Formation

Kireev, D. - Presenter, University of North Carolina At Chapel Hill

Hathaway, N., The UNC Eshelman School of Pharmacy

Williams, M., University of North Carolina at Chapel Hill

Xiaokang, Y., Department of Chemical and Biological Engineering, Princeton University,

Heterochromatin is a physical state of the chromatin fiber that maintains gene repression during cell development. Although evidence exists on molecular mechanisms involved in heterochromatin formation, a detailed structural mechanism of heterochromatin formation needs a better understanding. We made use of a simple simulation model with explicit representation of key molecular events, to observe molecular self-organization leading to heterochromatin formation. This simple unbiased simulation provides a comprehensive structural interpretation of micro- and macroscopic heterochromatin properties. In particular, this study suggests a structural mechanism by which chromatin bridging potentially promotes formation of condensed chromatin. It also elucidates the structural roots of a yet poorly understood phenomenon of a non-deterministic nature of heterochromatin formation and subsequent gene repression. Experimental Chromatin *in vivo* Assay was used to corroborate computational results with the dynamics of the repressive response to a heterochromatin-triggering event.

Phase-Separated Heterochromatin Domains Impart Mechanical Stiffness to the Nucleus

Surovtsev, I. - Presenter

Mochrie, S. G. J., Yale University

Williams, J., Yale University

King, M., Yale University

Nguyen, H., Yale University

Schreiner, S., Yale University

Hu, Y., Yale University

Heterochromatin association with the nuclear periphery is assumed to reflect its transcriptional silencing. However, recent work in mouse rod cells showed that peripheral localization is not required for gene silencing. Moreover, studies from our and other labs revealed that chromatin regions associated with the inner nuclear membrane play an integral role in the mechanical response of the nucleus. Since heterochromatin regions are spatially organized through liquid-liquid phase separation (LLPS), we asked whether phase-separated chromatin domains may have distinct mechanical properties and contribute to the emergent mechanical properties of the nucleus.

We use genetically-tractable *S. pombe* model to examine how altering heterochromatin formation or composition influences nuclear mechanics. We leverage image reconstruction software to quantitatively measure nuclear envelope fluctuations, a novel force spectroscopy that employs optical tweezers to directly measure viscoelastic properties of isolated nuclei, and microscopy approaches to monitor the dynamics of heterochromatin foci in living cells.

We find that both disrupting histone H3K9 methylation and loss of the histone H3K9me2/3 binding protein Swi6, an HP1 orthologue known to undergo LLPS, compromises nuclear stiffness. Accordingly, loss of the H3K9 demethylase Epe1 that leads to heterochromatin spreading increases nuclear stiffness. Mechanical response of the nucleus is also altered in cells expressing an allele of Swi6 with a mutation in its N-terminal domain critical for phase separation. This mutated protein is defective in transcriptional silencing, but not in dimerization or histone binding, suggesting that this Swi6 allele may uncouple the phase-separating properties of Swi6 from its other biochemical activities. Indeed, we found that cells expressing this Swi6 allele demonstrate altered accumulation of Swi6 in heterochromatic foci and defects in the coalescence of individual heterochromatin domains.

Altogether, these results argue that, beyond transcriptional effects, altering histone modifications has mechanical consequences. More generally, our work highlights that phase separated domains can perform mechanical work.

Submolecular-Resolution 3D Dynamics of Heterochromatin Formation

Kireev, D. - Presenter, University of North Carolina At Chapel Hill

Hathaway, N., The UNC Eshelman School of Pharmacy

Xiaokang, Y., Department of Chemical and Biological Engineering, Princeton University,

Williams, M., University of North Carolina at Chapel Hill

Heterochromatin is a physical state of the chromatin fiber that maintains gene repression during cell development. Although evidence exists on molecular mechanisms involved in heterochromatin formation, a detailed structural mechanism of heterochromatin formation needs a better understanding. We made use of a simple simulation model with explicit representation of key molecular events, to observe molecular self-organization leading to heterochromatin formation. This simple unbiased simulation provides a comprehensive structural interpretation of micro- and macroscopic heterochromatin properties. In particular, this study suggests a structural mechanism by which chromatin bridging potentially promotes formation of condensed chromatin. It also elucidates the structural roots of a yet poorly understood phenomenon of a non-deterministic nature of heterochromatin formation and subsequent gene repression. Experimental Chromatin *in vivo* Assay provided an unbiased estimate of time scale of repressive response to a heterochromatin-triggering event.

A Bisulfite-Free Approach for Base-Resolution Analysis of Genomic 5-Carboxylcytosine

Licyte, J. - Presenter, Vilnius University, Life Science Center

Gibas, P., Institute of Biotechnology, Vilnius University

Skardži?t?, K., Institute of Biotechnology, Vilnius University

Stankevi?ius, V., Institute of Biotechnology, Life Sciences Center, Vilnius University

Rukš?nait?, A., Institute of Biotechnology, Life Sciences Center, Vilnius University

Kriukien?, E., Institute of Biotechnology, Vilnius University

Due to an extreme rarity of 5-carboxylcytosine (5caC) in the mammalian genome, investigation of its role brings a considerable challenge. Methods based on bisulfite sequencing have been proposed for genome-wide 5caC analysis. However, bisulfite-based sequencing of scarcely abundant 5caC demands significant experimental and computational resources, increasing sequencing cost. Here, we present a bisulfite-free approach, caCLEAR, for high-resolution mapping of 5caCGs. The method uses an atypical activity of the methyltransferase eM.SssI to remove a carboxyl group from 5caC, generating unmodified CGs, which are localized by uTOP-seq sequencing. Validation of caCLEAR on model DNA systems and mouse ESCs supports the suitability of caCLEAR for analysis of 5caCGs. The 5caCG profiles of naive and primed pluripotent ESCs reflect their distinct demethylation dynamics and demonstrate an association of 5caC with gene expression. Generally, we demonstrate that caCLEAR is a robust economical approach that could help provide deeper insights into biological roles of 5caC.

Heterogenous Global Demethylation in Naïve Embryonic Stem Cells Results from Differences in DNA Methylation Maintenance in Single Cells

Chialastri, A. - Presenter, University of California Santa Barbara

Dey, S., University of California, Santa Barbara

Epigenetic remodeling, including genome wide erasure of 5-methylcytosine (5mC) is associated with the acquisition of pluripotency. To achieve this, cells can exhibit passive demethylation, where 5mC is not faithfully copied to newly synthesized DNA during replication. Recently, we have shown that the genome wide erasure of 5mC in early mouse and human embryogenesis is heterogenous, with a subset of cells experiencing global passive demethylation[1]. These findings indicate that tools to investigate passive demethylation of 5mC are critical in understanding pluripotency and development. Here we describe CpG-Dyad-sequencing (CpG-Dyad-seq) which combines enzymatic detection of modified cytosines and traditional nucleobase conversion techniques to identify the presence of hemimethylation or hemihydroxymethylation at the resolution of a single CpG dyad site. We then scale this technique down and allow for the simultaneous detection of RNA transcripts all from the same single cell. By applying single-cell CpG-Dyad-seq to different time points of mouse embryonic stem cells transitioning from a primed to a naÃ¯ve state of pluripotency, we observe extreme demethylation dominated by passive processes and discover this process is highly heterogenous and delayed in some cells. By connecting RNA expression from the same cells, we detect a small set of genes directly linked to 5mC levels during this transition. Finally, we determine that regions of the genome which escape 5mC reprograming do so by retaining high levels of 5mC maintenance and are associated with specific histone modifications.

1. Sen, M. *et al.* Strand-specific single-cell methylomics reveals distinct modes of DNA demethylation dynamics during early mammalian development. *Nat. Commun.* **12**, 1286 (2021).

Targeted Nuclease Strategies for Mapping R-Loop Chromatin Structures.

Sarma, K. - Presenter, The Wistar Institute

R-loops are three stranded nucleic acid structures with essential roles in many nuclear processes. However, their unchecked accumulation as seen in some neurodevelopmental diseases and cancers and is associated with compromised genome stability. Genome-wide profiling of R-loops in normal cells and their comparison in disease states can help identify precise locations of pathogenic R-loops and advance efforts to attenuate deviant R-loops while preserving biologically important ones. Toward this, we have developed an antibody-independent R-loop detection strategy, BisMapR, that combines nuclease-based R-loop isolation with non-denaturing bisulfite chemistry to produce high-resolution, genome-wide R-loop profiles that retain strand information. Furthermore, BisMapR achieves greater resolution and is faster than existing strand-specific R-loop profiling strategies. We applied BisMapR to reveal discrete R-loop behavior at gene promoters and enhancers. We show that gene promoters exhibiting antisense transcription form R-loops in both directions. and uncover a subset of active enhancers that, despite being bidirectionally transcribed, form R-loops exclusively on one strand. Thus, BisMapR reveals a previously unnoticed feature of active enhancers and provides a tool to systematically examine their mechanisms in gene expression.

Using Single Cell, Noncoding CRISPR/Cas9-Based Regulatory Element Screening to Dissect Regulatory Mechanisms in Complex Genetic Loci

Bounds, L. - Presenter, Duke

Gersbach, C. A., Duke University

The Major Histocompatibility (MHC) Locus is the most SNP-dense region in the human genome and has been linked with >100 polygenic disorders. However, it remains unknown which, if any, of those variants disrupt regulatory interactions and drive specific phenotypes. To dissect regulatory mechanisms within this locus, we coupled CRISPR/Cas9-based regulatory element screening with single cell RNA-sequencing to identify putative regulatory element (pRE)-gene pairs in three diverse cell types: human induced pluripotent stem cells (hiPSCs), hiPSC-derived neural progenitor cells (hNPCs), and the immortalized myelogenous leukemia K562 cell line.

We designed a library (n=12,723 sgRNAs) targeting a union set of DNase hypersensitivity sites (DHSs) and previously identified POU5F1 enhancers (n = 581 elements) with 150 TSS-targeting controls and 1,273 nontargeting controls. We transduced hiPSCs, hNPCs, and K562s, expressing dCas9-KRAB, and profiled 250,000, 340,000, and 120,000 single cell transcriptomes, respectively, seven to nine days post-transduction. We then performed differential expression testing to identify sgRNA-gene pairs. Using a threshold based on significance and effect size, we recover >170 DHS-gene pairs in each cell type and find the same pREs can regulate different genes in different cell types, and certain genes are only regulated by the targeted elements in one cell type. In all cell types, we show that pREs regulate target genes nearby (<10kb) and further away (>0.5 Mb) and are located in promoters, introns, 3â€™ UTRs, and distal intergenic regions. We show that distal and promoter pRE regions have distinct characteristics. In hiPSCs and hNPCs, promoter pREs are enriched for active and repressive histone marks, suggesting that poised promoters may have additional regulatory function. We also demonstrate that targeted enrichment of the gene expression libraries enables detection of regulatory elements for lowly expressed genes.

Epigenetic Regulation of the Kcc2/KCC2 Promoter for Gene Expression Enhancement in Pathologic Pain and Itch

Liedtke, W. - Presenter, Regeneron Pharmaceuticals

Excitatory-inhibitory balance is vital for physiologic functioning of the central nervous system. For incoming sensory information, various forms of injury to neural structures can cause pain. Pain establishing itself in the central nervous system in terms of chronic, pathologic pain is associated with less effective action of the neurotransmitters GABA and glycine. These transmitters are critical for normal relay in the sensory gate of the spinal cord, the spinal cord dorsal horn. Here, sensory relay neurons are affected by elevated intra-neuronal chloride levels which in turn corrupts inhibitory neurotransmission by GABA and glycine. The cause for elevated intra-neuronal chloride is lack of gene expression of the *Kcc2/KCC2* gene (mouse/human gene), caused by the neural injury. Because of relative absence of the *Kcc2/KCC2* gene product, the neuronal chloride extruding transporter, KCC2, chloride ions are not extruded sufficiently out of these sensory relay neurons, thus increasing the excitability of this primary afferent neural circuit. In order to remedy this particular aspect of the etio-pathogenesis of chronic pathologic pain, we have focused on epigenetic regulation of the *Kcc2/KCC2* promoter activity and conducted a molecular screening campaign with the goal of finding *Kcc2/KCC2* gene expression-enhancing compounds. Out of 2 NCI chemical libraries of shelved cancer drugs (n=1057), we found 137 first-round hit compounds. These 137 winners were interrogated more in-depth with a yield of 4 top-winner compounds, of which we selected kinase inhibitor Kenpaullone. We demonstrated Kenpaullone to inhibit GSK3-beta as a neuro-cellular mechanism to potently and lastingly attenuate pain of nerve constriction, inflammation and bone cancer, as well as chronic itch of contact allergy, in preclinical mouse models. GSK3-beta inhibition by Kenpaullone facilitated nuclear transfer of delta-catenin, which bound to KAISO transcription factors and DNA binding sites in the *Kcc2* promoter, switching *Kcc2* gene repression to enhancement via this specific epigenetic mechanism.

Institutionalizing Epigenetic Treatment with New Tools; Analysis, Diagnosis and Treatment Via the Enteric Nervous System Utilizing an Engineering Platform.

Hopkins, L. - Presenter

Case studies discussing the release of epigenetic potential in patient care can be found across the healthcare field. Some of the challenges associated with institutionalizing epigenetic based treatments include: Familiarity; practitioner jurisdiction disputes; as well as patient and practitioner bias and/or prejudice regarding treatment modalities.

The Hopkins Bioengineering Research Cooperative (HBRC) has spent over two decades researching and developing new engineering tools that are capable of eliminating these problems. With these tools, growth vector sets can be analyzed, diagnosed and released via the enteric nervous system.

Enteric Structural Mechanics, Enteric Axial Harmonics, and The Athena Series introduce a new territory within chemical engineering and mathematics. Enabling engineers to map a set of thermodynamic equations, these tools can measure and quantify epigenetic potentials in biological systems using reactor design concepts and process control strategies. Defined as the Primary Economy, the energy available to support biological infrastructure within human community systems can now be understood as a reserve: Analysis, diagnosis, and treatment protocols can be administered utilizing an engineering platform to optimize growth vector release and distribution in clinical application.

In response to a federal review, HBRC has requested government assistance to establish a research and development clinic where these engineering tools can be applied within a community setting with participation from all sectors: Residential, government, academic, and business. Here, collective resources can be pooled and shared as teaching and training programs are established to institutionalize a new paradigm for the standardization of epigenetic based therapies to manage the Primary Economy.

Observing Neurodegeneration in Real Time Using 2-Photon Microscopy in the LSD1 Inducible Mouse Model

Gosrani, S. - Presenter

Katz, D., Emory University

AD is characterized by two hallmark pathologies: the accumulation of pathological AÎ² plaques and neurofibrillary tau tangles (NFTs). However, the role of NFTs in mediating neurodegeneration remains unclear. We have previously demonstrated that the H3K4 histone demethylase LSD1/KDM1A is mislocalized to cytoplasmic NFTs in AD cases. In agreement with this, we discovered that hippocampal deletion of LSD1 resulted in widespread loss of pyramidal neurons, a prominent feature of AD. Additionally, the neurodegeneration caused by loss of LSD1 is associated with AD-specific gene expression changes and learning and memory deficits, suggesting that the neurodegeneration in the LSD1 mouse model may be related to AD patients. In order to better understand neuronal cell death mechanisms, we propose using two-photon live imaging in the LSD1 inducible knockout mouse to visualize neurons dying in the cerebral cortex. To visualize neurons dying in the inducible LSD1 mouse model, we incorporated a Thy1-YFP neuronal reporter. Using a two-photon microscope, we will be able to visualize YFP-expressing neurons in an awake animal through a surgically implanted cranial window. Given the robust neurodegeneration phenotype, loss of LSD1 will generate a feasible and rapid timescale to view neurons degenerating in real time. Previously, we have shown that axons and dendrites are completely lost from hippocampal neurons in the LSD1 inducible knockout mice. Using live imaging, we hope to determine whether this is an active part of neuronal cell death in this model. In addition, the loss of LSD1 induces the inflammatory response, suggesting microglia may play a role in neurodegeneration. We will interrogate the role of the microglia activation in this system by modulating its function. Collectively, these studies will provide insight into how inhibition of the chromatin repressor LSD1 contributes to neurodegeneration, with the ultimate goal of generating novel therapeutic strategies targeting LSD1.

The Inhibition of LSD1 VIA Sequestration Contributes to Tau-Mediated Neurodegeneration

Bai, Y. - Presenter, Emory University

Katz, D., Emory University

Alzheimerâ€™s disease (AD) is characterized by the abnormal aggregation of Î²-amyloid plaques and neurofibrillary tangles of hyperphosphorylated tau (NFTs). However, the molecular mechanism underlying the neuronal cell death is not well understood. The H3K4me1/2 demethylase, LSD1 functions in nucleus to repress genes. Surprisingly, our lab found that LSD1 is mis-localized with cytoplasmic pathological tau in human AD cases, that suggests LSD1 may be inhibited by tau. To study the function of LSD1, our lab showed that the inducible deletion of LSD1 in adult mice induces cortical and hippocampal neurodegeneration, learning and memory deficits, and transcription alternations that match human AD cases. Therefore, we hypothesized that pathological tau contributes to neuronal cell death by sequestering LSD1 into the cytoplasm.

To address this hypothesis, we utilized the PS19 tauopathy mouse model to examine the functional consequences of changing LSD1 expression and to examine the interaction between pathological tau and LSD1. We found that LSD1 is depleted from the nucleus and mis-localized with cytoplasmic NFTs in adult PS19 Tau mice, supporting the hypothesis that NFTs sequester LSD1. In addition, we found that changing LSD1 expression can specifically modulate tau-mediated neurodegeneration in PS19 mice. This suggests that pathological tau is functioning in part through LSD1. Finally, we found that LSD1 interacts with pathological tau through its N-terminus.

We conclude that pathological tau functions in part by sequestering LSD1 in the cytoplasm and preventing it from continually blocking inappropriate gene expression. A model for the pathological interaction between LSD1 and tau will be presented.

Nuclear Organisation

M, V. - Presenter, AMRITA COLLEGE OF ENGINEERING

Nuclear organization

Nuclear organization refers to the spatial distribution of chromatin within a cell nucleus. There are many different levels and scales of nuclear organization. Chromatin is a higher order structure of DNA.

Levels of nuclear organization

Linear DNA and chromosome basics

The first level of genome organization concerns how DNA is arranged linearly, and how it is packaged into chromosomes. DNA is composed of two antiparallel strands of nucleic acids, with two bound and opposing nucleic acids referred to as DNA base pairs.

Chromosomal domains

Self-interacting domains

Self-interacting (or self-associating) domains are found in many organisms â€" in bacteria, they are referred to as Chromosomal Interacting Domains , whereas in mammalian cells, they are called Topologically Associating Domains (TADs). Self-interacting domains can range from the 1â€"2 mb scale in larger organisms to 10s of kb in single celled organisms.

Lamina-associating domains and nucleolar-associating domains

Lamina-associating domains (LADs) and nucleolar-associating domains (NADs) are regions of the chromosome that interact with the nuclear lamina and nucleolus, respectively. Making up approximately 40% of the genome, LADs consist mostly of gene poor regions and span between 40kb to 30Mb in size.

A/B compartments

A/B compartment-associated regions are on the multi-Mb scale and correlate with either open and expression-active chromatin ("A" compartments) or closed and expression-inactive chromatin ("B" compartments).A compartments tend to be gene-rich, have high GC-content, contain histone markers for active transcription, and usually displace the interior of the nucleus. B compartments, on the other hand, tend to be gene-poor, compact, contain histone markers for gene silencing, and lie on the nuclear periphery.

Chromosome territories

The last level of organization concerns the distinct positioning of individual chromosomes within the nucleus. The region occupied by a chromosome is called a chromosome territory (CT). Among eukaryotes, CTs have several common properties.

Simultaneous Measurement of DNA Methylation and Genome-Nuclear Lamina Interactions in Single Cells

Podorefsky, D. - Presenter, UC Santa Barbara

Dey, S., University of California, Santa Barbara

Chialastri, A., University of California Santa Barbara

Smith, M., UC Santa Barbara

DNA methylation (5-methylcytosine or 5mC) and the 3-dimensional organization of the genome within the cell nucleus are two critical epigenetic features that regulate gene expression and cellular behavior, and aberrant patterns of these epigenetic features are associated with cancer and disease [1-3]. In cancer, widespread regions of DNA methylation loss, termed hypomethylation, have previously been found positioned at the periphery of the nucleus, the nuclear lamina (NL), suggesting DNA methylation and genome organization could be linked epigenetic states [1]. However, the direct genome-wide relationship between 5mC and genome-NL interactions remains obscured in bulk measurements. To overcome this limitation, we developed a new single-cell sequencing technology (sc5mC+DamID-seq) to simultaneously measure DNA methylation and genome-NL interactions from the same cell. sc5mC+DamID-seq uses barcoded adapters unique to each epigenetic feature and individual cell, enabling us to profile thousands of single cells per day. By applying sc5mC+DamID-seq to chronic myelogenous leukemia cells (KBM7), we observed hypomethylation at regions contacting the NL, known as lamina associated domains (LADs). Interestingly, we also discovered that genomic regions that contact the NL more frequently in single cells display the greatest loss and variability in 5mC. Further, while LADs appear as continuous stretches of contact with the NL in bulk sequencing, LADs at the single-cell level frequently display segments of noncontact with the NL, which we found directly correlate with mean levels of 5mC along each LAD. Finally, to test the connection between these two epigenetic features, we globally demethylated the epigenome, which relocated more variable genome-NL contact regions towards the nuclear interior. Thus, simultaneous single-cell measurements in sc5mC+DamID-seq has enabled us to systematically uncover the relationship between DNA methylation and genome organization in cancer cells.

References:

[1] Berman, B.P. *et al. Nat. Genet.* **2012**,44,40-46.

[2] Pakneshan, P. *et al. CCR* **2004**,10,3035-3041.

[3] Agrelo, R. *et al. JCO* **2005**,23,3940-3947.

Substrate Deformation Regulates DRM2-Mediated DNA Methylation in Plants

Leichter, S. - Presenter

DNA methylation is a major epigenetic mechanism critical for gene expression and genome stability. In plants, domains rearranged methyltransferase 2 (DRM2) preferentially mediates CHH (H = C, T, or A) methylation, a substrate specificity distinct from that of mammalian DNA methyltransferases. However, the underlying mechanism is unknown. Here, we report structure-function characterization of DRM2-mediated methylation. An arginine finger from the catalytic loop intercalates into the non-target strand of DNA through the minor groove, inducing large DNA deformation that affects the substrate preference of DRM2. The target recognition domain stabilizes the enlarged major groove via shape complementarity rather than base-specific interactions, permitting substrate diversity. The engineered DRM2 C397R mutation introduces base-specific contacts with the +2-flanking guanine, thereby shifting the substrate specificity of DRM2 toward CHG DNA. Together, this study uncovers DNA deformation as a mechanism in regulating the specificity of DRM2 toward diverse CHH substrates and illustrates methylome complexity in plants.

Chromatin-Mediated Alternative Splicing Regulates Reward Behavior.

Heller, E. A. - Presenter, University of Pennsylvania

Lombroso, S.

Xu, S. J., University of Pennsylvania

Neuronal alternative splicing is a key gene regulatory mechanism in the brain. However, the spliceosome machinery is insufficient to fully specify splicing complexity. In considering the role of the epigenome in activity-dependent alternative splicing, we and others find the histone modification H3K36me3 to be a putative splicing regulator. In this study, we found that mouse cocaine self-administration caused widespread differential alternative splicing, concomitant with the enrichment of H3K36me3 at differentially spliced junctions. Importantly, only targeted epigenetic editing can distinguish between a direct role of H3K36me3 in splicing and an indirect role via regulation of splice factor expression elsewhere on the genome. We targeted Srsf11, which was both alternatively spliced and H3K36me3 enriched in the brain following cocaine self-administration. Epigenetic editing of H3K36me3 at Srsf11 was sufficient to drive its alternative splicing and enhanced cocaine self-administration, establishing the direct causal relevance of H3K36me3 to alternative splicing of Srsf11 and to reward behavior.

Epigenetic Modulation of H4K20 Methylation at Mitotic Centromeres

Hahn, S. - Presenter

Herlihy, C., Worcester Polytechnic Institute

Hermance, N., Worcester Polytechnic Institute

Manning, A., Worcester Polytechnic Institute

Crowley, E., Worcester Polytechnic Institute

Dynamic regulation of chromatin structure is critical for several biological processes, including accurate chromosome segregation during mitosis. Histone posttranslational modifications that occur at centromeric heterochromatin have the capacity to change the formation and function of the centromere and, ultimately, how the kinetochore interacts with the mitotic spindle during cell division. Defects in this regulation corrupt the way attachments are made and can result in chromosome segregation errors. Here I characterize the role of the Suv420, a histone lysine methyltransferase, in regulating centromere structure and kinetochore composition. My data indicate centromeric Histone 4 Lysine 20 (H4K20) trimethylation limits Aurora B localization to the centromere and plays a role in accurate chromosome segregation.

Regulation of Embryonic Cell Fate Decision By Histone Methylation

Rodriguez, J. D. - Presenter

Katz, D., Emory University

In *C. elegans*, two epigenetic enzymes, the H3K4me2 demethylase, SPR-5, and the H3K9 methyltransferase, MET-2, are maternally deposited into the oocyte and cooperate to reestablish the epigenetic ground state by modifying histone methylation. Progeny of worms lacking *spr-5* and *met-2* accumulate high levels of H3K4me2 in the subsequent generation. This inappropriate H3K4me2 is associated with a low level of embryonic lethality, as well as a severe developmental delay and sterility in the small number of animals that reach adulthood. In addition, the progeny of *spr-5;met-2* mutants improperly express germline genes in somatic tissues. We hypothesized that this improper expression could result in cell lineage defects. To interrogate how reprogramming defects may affect early embryonic development, we are taking advantage of the invariant *C. elegans* embryonic lineage by performing automated lineage tracing experiments in *spr-5;met-2* progeny. Thus far, we observe three stages of defects. First, in early embryogenesis, the shape of the embryos is rounder, with a diameter that is greater than N2, and the mutant cells divide faster than N2. Since these defects are prior to the onset of zygotic transcription, they may be due directly to the decondensed state of chromatin associated with increased H3K4me2, or indirectly due to altered maternal RNAs and proteins. Second, after the onset of zygotic transcription, we observe many changes in cell timing, along with defects in cell migration. These defects may be due to the inappropriate somatic transcription of germline genes that we observe in the progeny of *spr-5;met-2* mutants. We are currently performing single-cell RNA sequencing in the progeny of *spr-5;met-2* mutants to further explore these three stages of defects.

Systematic Comparison of CRISPR-Based Transcriptional Activators Uncovers Gene-Regulatory Features of Enhancer-Promoter Interactions

Wang, K. - Presenter, Rice University

Li, J., Rice University

Mahata, B., Rice University

Goell, J., Rice University

Escobar, M., Rice University

Hilton, I., Rice University

Nuclease-inactivated CRISPR/Cas-based (dCas-based) systems have emerged as powerful technologies to synthetically reshape the human epigenome and tune gene expression. Despite the straightforward programmability and rapidly increasing adoption of these platforms, the relative potencies, and mechanistic differences between CRISPR/Cas-based transcriptional activators are incompletely characterized, particularly at human enhancer-promoter pairs. Here we systematically compared the transcriptional effects of the most potent and widely used dCas9-based transcriptional activators, as well as a novel activator consisting of dCas9 fused to the catalytic core of the human CBP protein, at human enhancer-promoter pairs. We find that although each system can generally transactivate targeted human loci, the platforms display variable expression levels in different human cell types and further, that the relative efficacies for transcriptional activation vary substantially based upon the effector domain used, recruitment architecture, targeted locus, and the cell type assayed. We also show that when targeted to an enhancer, each dCas9-based activator can induce the production of enhancer RNAs (eRNAs), and that this induction of eRNAs is tightly correlated with downstream mRNA expression from a cognate promoter. Our studies also reveal that different dCas9-based transcriptional activators can deposit histone acetylation at a targeted enhancer, which results in a highly correlated increase in histone acetylation at a downstream promoter. Further, we use dCas9-based transcriptional activators to show that an intrinsic transcriptional and epigenetic reciprocity exists between a testbed human enhancer-promoter pair. Finally, our findings demonstrate that contact frequencies between an enhancer-promoter can highly correlate to transcriptional outputs, which can be manipulated using targeted dCas9-based transcriptional activators also suggest that. Collectively, our study provides new insights into the enhancer-mediated control of gene expression and the use of dCas9-based transcriptional activators at human regulatory elements.

An in Vitro Human Platform to Efficiently Study Angelman Syndrome Class I/II Deletion Genes

Authors:

Yagci, Z. B. - Presenter, North Carolina State University

Drobna, Z., North Carolina State University

Keung, A. J., North Carolina State University

Liu, H., The University of North Carolina at Chapel Hill

Simon, J., The University of North Carolina at Chapel Hill

Sen, D., North Carolina State University

Stuppy, S., North Carolina State University

Angelman Syndrome (AS) is a neurodevelopmental disorder characterized by mental disability, speech impairment, and ataxia. Deletions and loss-of-function mutations at the maternal allele of chr15q11-q13 can lead to AS. Most AS individuals have Class I/II deletions, large cytogenetic deletions spanning at least 5 Mb. Among the genes deleted within this region, *UBE3A* has been reported to be responsible for the core symptoms of AS. However, *UBE3A* and its surrounding genes constitutively contribute to the AS etiology. Hence, we need human-specific platforms that will enable us to investigate *UBE3A* and its understudied neighboring genes. The aim of this study is to establish an isogenic AS platform derived from human pluripotent stem cells which will help us investigate the roles of deleted or mutated genes in AS Class I/II deletion. Accordingly, we pursue: (i) Designing and integrating a recombinase-based landing pad into a safe harbor locus of AS patient-derived cell lines to rescue the expression of Class I/II deletion genes, (ii) Generating AS Class I deletion cell lines from human embryonic stem cells, (iii) Integrating a recombinase-based landing pad into the deletion region of an H9 maternal *UBE3A* deletion cell line to analyze *UBE3A* mutations. We successfully designed a landing pad and integrated it into the AAVS1 safe harbor locus of H9 cells and are in the process of integrating it into AS deletion cell lines. In parallel, we are using CRISPR/Cas9 to induce large cytogenetic deletions in H9 and H1 cells. Our work on deletion cell lines showed that there were differences in gene orders between genome assemblies for the AS deletion region. To shed light on this, we have obtained haplotype aware, phased genome sequences of H1 and H9 cells.

Bidirectional Post-Mitotic Reprogramming of H3K27me3 Underlies Neuronal Maturation and Plasticity in the Cerebellum

Ramesh, V. - Presenter, Duke University

Liu, F., Duke University

Minto, M., Duke University

Chan, U., Duke University

West, A., Duke University

Terminally differentiated cells including neurons dynamically regulate their genome despite leaving the cell-cycle early in life in order to assimilate new information throughout adulthood. The molecular basis of this dynamic gene regulation remains poorly understood yet when disrupted contributes to defects in synaptic plasticity and cognition. Work from our lab has shown the dynamic, post-mitotic redistribution of the repressive histone modification H3K27me3 in a locus-specific fashion across the genome of the maturing cerebellar granule neuron (CGN). We are interested in asking how a fully fate committed neuron marks particular loci for differential histone methylation, how is this controlled in maturing neurons and what is the biological consequence of this process.

We are using the maturation of CGNs in the murine cerebellum as a model system to study post-mitotic chromatin regulation. CGN-precursors terminally differentiate postnatally and undergo several timed gene expression events in order to migrate, populate the innermost layer of the cerebellar cortex, and undergo dendrite outgrowth and synaptogenesis. We are utilizing Chromatin Immunoprecipitation to ask how CGNs in culture and in vivo redistribute H3K27me3 in response to genetic, pharmacological, and physiological perturbations. Additionally, we use CRISPR-based epigenome editing methods to ask whether this modification causally acts to regulate gene expression.

We find that the activity of H3K27me3 regulators EZH2 and KDM6B temporally regulates genes essential for cerebellar maturation. EZH2-mediated H3K27me3 is required for expression of genes that signify commitment to the neuronal fate, whereas KDM6B-mediated H3K27 demethylation promotes the expression of a set of late expressed CGN genes that are associated with maturation of synaptic function. Taken together these data show that dynamic regulation of the repressive chromatin state afforded by H3K27me3 is important for timing programs for gene expression in maturing postmitotic neurons long after closure of the period of fate commitment.

Characterization of a Potential Gene Interaction between Chromatin Modifiers Spr-5, Met-2, and Mep-1 in Determining Germline Versus Soma in C. Elegans

Authors:

Chavez, S. - Presenter, Emory University

Katz, D., Emory University

Brockett, J., Emory University

Carpenter, B., Kennesaw State University

Schmeichel, K., Oglethorpe University

Birol, O., Georgia Institute of Technology

In *C. elegans*, histone modifying enzymes aid in the activation and repression of genes that are required to distinguish germline from soma. For example, SPR-5 is a histone demethylase that removes the activating histone modification H3K4me1/2 and MET-2 is a histone methyltransferase that adds the repressive histone modification H3K9me2. Together, these chromatin modifiers work synergistically to establish a totipotent ground state by shutting down the transcription of germline specific genes that were expressed in the germline of the parents. Previous studies have shown that *spr-5;met-2* double mutants produce progeny that experience a severe developmental delay. Transcriptomic analysis indicated that the delay is largely due to inappropriate germline gene expression in somatic tissues. Here we explore the possibility that the chromatin state established in the early embryo by SPR-5 and MET-2 is reinforced in the later embryo by ATP-dependent chromatin remodeling activity. MEP-1 is a component of the ATP-dependent epigenetic deacetylase complex, MEC. Similar to *spr-5;met-2* double mutants, *mep-1* mutants fail to suppress germline genes in somatic tissues, as indicated by the misexpression germline proteins in the soma of L1 larvae. We tested a potential interaction between these pathways by subjecting *spr-5*, *met-2* and *spr-5;met-2* mutants to *mep-1* RNAi. *mep-1* knockdown results in an exacerbated developmental delay phenotype in these mutant backgrounds. The â€œtripleâ€ mutant is most severe, displaying a full L1 arrest. To fully understand the molecular basis of this synergy, we performed RNA-sequencing. We find very little evidence of new expression changes induced in the triple mutants compared to controls. Instead, we find that expression changes in the *spr-5;met-2* double mutants are exacerbated in triple mutants. This suggests that MEP-1 specifically reinforces the repressed chromatin state established by SPR-5 and MET-2 reprogramming at fertilization. These results help establish how chromatin modifying enzymes work together to establish developmental cell fates.

Enhancer-Mediated Regulation of Bdnf

Narayanan, A. - Presenter

Brain Derived Neurotrophic Factor (BDNF) plays essential roles in neuronal survival, differentiation, and plasticity, and it contributes to learning and memory. The expression of BDNF is highly induced by neuronal activity and is mediated primarily at the transcriptional level, via the regulation of stimulus-modulated transcription factors. Although much is known about the mechanisms of *Bdnf* regulation at this geneâ€™s promoters, we do not know about the presence and contribution of distal enhancers in regulating activity-dependent *Bdnf* expression. To address this, we first aligned a 5C-seq dataset to ChIP datasets for many biochemical features present at active enhancers. From this analysis, we identified six regions that show strong indications of being active enhancers upon neuronal stimulation. To functionally test the role of these specific putative gene regulatory elements in the regulation of *Bdnf* induction, we used dCas9-KRAB based CRISPRi tools in embryonic mouse cortical neuron cultures stimulated by KCl. When dCas9-KRAB was recruited to *Bdnf* promoters, more than a two-fold reduction in *Bdnf* expression was seen. However, when dCas9-KRAB was recruited to the putative enhancers, a weaker suppression in *Bdnf* expression was seen for two regions. RNA-seq data from these regions show low levels of eRNA expression upon neuronal activation, suggesting that these regions may not be enhancers. As it is possible that these regions may be stronger enhancers under different stimulation conditions, in the future we will interrogate these regions upon bicuculline stimulation. Also, since this suggests that eRNA may be a more robust marker for active enhancers, we will be using RNA-seq datasets and eRNA quantification methods to narrow in and identify *Bdnf* enhancers. However, with the data collected from the experiments thus far, it is still unclear if these regions are indeed distal *Bdnf* enhancers.

Epigenetic Modification Sensing

Jumbo, T. - Presenter, University of PORTHARCOURT

SENSING EPIGENETIC MODIFICATION

Tamunotonye Jumbo â€“ B.Eng. (in view)

Department of Chemical Engineering, University of Portharcourt, Choba Nigeria.

Abstract

Epigenome is the data set of all the changes that occurs in the DNA and to the DNA associated protein (histone).

Epigenetic modifications are restorable alterations in the chemical or physical changes in the substance consisting of DNA and protein within a chromosome. This substance is known as chromatin. There are two main types of epigenetic modifications:

- DNA methylation â€“ Adding Methyl to the DNA strings.
- Histone tails alteration â€“ Alteration in DNA associated protein.

Epigenetic modifications causes various forms of diseases and health conditions which includes: lupus, asthma, diabetes, rett syndrome, Î±-thalassemia, X-linked syndrome, Alzheimerâ€™s disease, Rubinstein-Taybi syndrome, Parkinsonâ€™s disease and Coffin-Lowry syndrome, cancer and neurological disorders. However, epigenetic modifications could be detected by:

- **Pyrosequencing analysis:** this provides overview of the methylation pattern and it also allows for the measurement of the methylation level of individual CpG dinucletodes. It analyzes any CpG site of interest.
- **Bisulfite conversion**: In this method, a DNA sample is treated with sodium bisulfite resulting in the deamination of unmethylated cytosine to uracil and allowing the distinction between cytosine and methylated cytosine via sequencing.
- **DNADNA enzyme digest:**This method is based on the use of DNA endonucleases which do not cut methylated DNA.
- **Identify modified-histone interacting proteins**: Immunoprecipitation and WB analyses are used in combination for the identification of histone interacting proteins. Co-immunoprecipitation of histone-binding protein complexes requires the enzymatic processing of chromatin to generate single nucleosome units.
- **Detect histone modifications:** Immunoblot analysis including dot blots and WB are used to validate antibodies specificity or to access histone mark plentifulness.
- **Determine the genomic location of histone marks**: Chromatin immunoprecipitation (ChIP) is used in combination with PCR, next generation sequencing (NGS).

Histone Modification during Mouse Embryo Development

Authors:

Taylor, J. - Presenter, UoE

Harasani, A. - Presenter, UoE

Pennings, S., UoE

Epigenetic mechanisms such as DNA methylation and histone modifications shape the embryonic epigenome during development. Early embryos are marked by H3 trimethylation on lysine 27 (H3K27me3), a repressive epigenetic marker. It is facilitated by the Polycomb group repressive complex 2 (PRC2) subunit of (EZH1/2), an enzyme that epigenetically regulates chromatin structure and gene expression by engaging DNA methyltransferases with H3K27me3-mediated gene silencing. As a result, EZH2 is crucial for stem cell renewal, maintenance, and differentiation into several cell lineages.

During the preimplantation phase of the embryo, the totipotent zygote undergoes continual cleavage divisions, resulting in the blastocyst. The blastomeres forming the trophectoderm (TE), epithelium that gives rise to placental cells, and the inner cell mass (ICM) that will form the embryo proper are separated at the blastocyst stage. The ICM divides into the epiblast (EPI) and primitive endoderm (PE) before embryo implantation (PrE). This cell lineage specification is assumed to be regulated by transcription factors and epigenetic mechanisms, which govern gene expression in place and time without impacting the genetic code.

In this study, two small molecule inhibitors of H3K27me3 methylase (EZH2), GSK-126 and EPZ-6438, were used to lower H3K27me3 levels and activity in preimplantation mouse embryos in vitro. Mouse embryos were collected at the E0.5 stage and cultured in the presence of GSK-126, EPZ-6438, or DMSO until the E4.5 stage. After that, immunocytochemistry and qPCR were used to evaluate the effect of EZH2 inhibition on expression. The findings were analysed to see if EZH2-mediated gene repression may impact cell fate and proliferation in preimplantation mouse embryos.

HP1 Oligomerization Compensates for Low-Affinity H3K9me Recognition and Provides a Tunable Mechanism for Heterochromatin-Specific Localization

Ragunathan, K. - Presenter, University of Michigan

HP1 proteins bind with low affinity but high specificity to histone H3 lysine 9 methylation (H3K9me), forming transcriptionally inactive genomic compartments referred to as heterochromatin. How HP1 proteins traverse a complex and crowded chromatin landscape on the millisecond timescale and yet recognize H3K9me with high specificity remains paradoxical. Here, we visualize the single-molecule dynamics of an HP1 homolog, the fission yeast Swi6, in its native chromatin environment. By analyzing the motions of individual Swi6 molecules, we identify mobility states that map to discrete biochemical intermediates. Using mutants that perturb Swi6 H3K9me recognition, oligomerization, or nucleic acid binding, we parse how each biochemical property affects protein dynamics in living cells. Based on *in vivo* kinetic measurements, we estimate that Swi6 recognizes H3K9me in cells with ~94-fold specificity. While nucleic acid binding suppresses Swi6 oligomerization, as few as four tandem chromodomains are sufficient to overcome these inhibitory effects to facilitate Swi6 localization at sites of heterochromatin formation. Our studies propose that HP1 oligomerization is essential to form dynamic, higher-order complexes that can outcompete the inhibitory effects of nucleic acid binding to enable highly specific H3K9me recognition in living cells.

The Expression of Key Hub Genes within the Nucleus Accumbens Mediates Resilience to Chronic Stress.

Authors:

Gyles, T. - Presenter

Nestler, E., Icahn School of Medicine at Mount Sinai

Parise, E., Icahn School of Medicine at Mount Sinai

Godino, A., Icahn School of Medicine at Mount Sinai

Browne, C., Icahn School of Medicine at Mount Sinai

Major depressive disorder (MDD) is the leading leading contributor to suicide according to the World Health Organization. Half of all patients diagnosed with MDD are at least partly resistant to antidepressant treatment. Notably, both genetic and environmental factors, including exposure to chronic social stress, play a role in the development of MDD and related syndromes. The chronic social defeat stress (CSDS) paradigm in mice has proven to be a highly useful animal model for studying depression-related behavioral abnormalities. Importantly, this paradigm allows for the identification of animals that succumb to the effects of the stress, termed susceptible, from those that do not, termed resilient. To better understand the potential genes underlying the resilient phenotype, we performed RNA-sequencing on mice exposed to CSDS and compared gene expression changes within the nucleus accumbens (NAc) between resilient and susceptible mice. Using weighted gene co-expression network analysis (WGCNA), we identified gene expression patterns that are associated with resilience. Then using an undirected key driver analysis that allows for the reconstruction of accurate cellular networks (ARACNE) based on gene-gene correlations. We identified three genes (Gprin1, Bcr, and Stx1a) that were predicted to be key drivers (regulate other genes within a highly significant), resilient-specific module. We are now

investigating how changes in the expression of these genes could lead to changes in depressive-like behavior. To test this, multiple cohorts of mice were exposed to CSDS or another stress

paradigm, chronic variable stress, and mRNA expression levels of these target genes were assessed. We also tested the functional consequences of virally manipulating these genes following prior to CSDS. These studies suggests that overexpression of these genes in NAc neurons produces a pro-resilient effect. Together, these studies are providing novel insight into the molecular basis of stress resilience.

Thermodynamics and Kinetics of Blunt-End Driven DNA Origami Tiles Dimerization through Coarse-Grained Molecular Dynamics Simulations

Sharma, P. - Presenter, Duke University

Arya, G., University of California San Diego

Sensale, S.

Directed self-assembly of DNA origami structures offers a powerful way to create complex devices at the nanoscale, with promising applications in a broad range of fields such as biosensing, nanoscale robotics, photonics, computing, and drug delivery. However, the complexity of these structures obscures the fundamental principles of their assembly process. To unveil such principles, we developed a coarse-grained model for DNA origami structures which allows for kinetic and thermodynamic predictions of blunt end driven dimerization. The simplicity of this model, which accurately captures multiple features observed in experiments, allows us to characterize how the geometric, elastic, and energetic properties of the blunt end motifs modify the enthalpy and entropy of dimerization, and highlights the role of entropic effects in the previously reported (albeit not fully understood) sub-linearity of the free energy of dimerization with respect to the number of active binding sites. This model also allows us to predict the effects of cytosine methylation and DNA mismatches on binding energy, enabling tunable molecular recognition with many potential uses in nanotechnology.

Using DNA Methylation Analysis As a Tool to Develop in Vitro Models for Aging Studies

Kelkar, G. - Presenter, North Carolina State University

Drobna, Z., North Carolina State University

Keung, A. J., North Carolina State University

Aging is the leading risk factor for neurodegenerative diseases. To develop better therapeutics and treatment strategies for these disorders, it is important to understand the aging process as well as to develop models of aged organisms, tissues, and cells. However, due to ethical reasons and long timescales, this is challenging in a human context. Human pluripotent stem cells (hPSCs) can be differentiated to neural tissues and cells but often reflect fetal states. Our aim is to develop an *in vitro* model for accelerated aging using hPSC-derived neurons and cerebral organoids. Prior work has investigated the use of various perturbations to induce neurotoxicity or aging related phenotypes like heterochromatin reorganization. Here we exposed human neurons and organoids to progerin overexpression and oxidative stress and assessed phenotypic changes relevant to aging. To assess induction of aging in as unbiased a manner as possible, without relying on proxy cellular phenotypes that could arise directly from the perturbations themselves, we will measure whole genome DNA methylation profiles (DNAm). Whole genome DNAm profiling has been used to develop age-predicting algorithms called Epigenetic Clocks. We plan to apply the novel Fetal Brain Clock (FBC), to predict the chronological age of our perturbed samples. In parallel, we are developing a comparative analysis of modified 2D neurons, organoids, and human brain tissues. These studies will contribute to developing a standard aging technique for *in vitro* models like cerebral organoids. Furthermore, this model may be used to examine biomarkers and molecular processes relevant to neurodegenerative disorders.

AUTHOR INDEX